Energy 233

悠悠球和风筝

The Yo-yo and the Kite

Gunter Pauli

[比]冈特·鲍利 著

[哥伦]凯瑟琳娜·巴赫 绘

贾龙智子 译

上海遠東出版社

特别感谢以下热心人士对童书工作的支持：

匡志强　方　芳　宋小华　解　东　厉　云　李　婧

刘　丹　熊彩虹　罗淑怡　旷　婉　杨　荣　刘学振

何圣霖　王必斗　潘林平　熊志强　廖清州　谭燕宁

王　征　白　纯　张林霞　寿颖慧　罗　佳　傅　俊

胡海朋　白永喆　韦小宏　李　杰　欧　亮

目录

Contents

黄昏的暮色中，一只印度狐蝠正在沿着马尔代夫一座岛屿的海岸线飞驰。他发现了一条在海岸附近，看上去很自信的鲸鲨，于是开始和鲸鲨聊了起来：

“我几乎每天都会从很远的地方看到你。终于能认识你真是太好啦。”

“谢谢你。你知道的，这里是我在这世上最喜欢的地方。自从我在这些环礁附近安顿下来后，我就再没想过离开。”

An Indian flying fox is racing along the coast of an island in the Maldives at dusk. He spots a whale shark close to shore, and feeling confident, he starts a conversation by saying:

"I see you from afar almost every day. How good to finally meet you."

"Thank you. This is my favourite spot in the world, you know. Since I have settled around these atolls, I've never felt like leaving."

一只印度狐蝠正在沿着海岸线飞驰

An Indian flying fox is racing along the coast

……拍自拍……

… taking selfies …

“好多好多人从世界各地飞来看你。今天早上至少有二十艘船只是为了看你一眼而驶过这里。”

“我必须承认那些游客有点讨厌。似乎他们唯一关心的就是以我为背景拍自拍。”

“So many people fly in from around the world to see you. At least twenty boats were racing across out there this morning to get a glimpse of you.”

“Those tourists are a bit of a nuisance, I must admit. The only thing they seem to care about is taking selfies, with me in the background.”

“唉，看起来没人想跟我合影！人们觉得我是害兽，而我做的只是吃掉了那些因为没人收获而过熟的果子……”

“成千上万的游客抵达我们这个小岛，真是让我担心。他们吃什么呢？空运过来的冷冻鸡肉吗？而且怎样才能满足他们对淡水以及让他们能享受凉爽冷气的电力的需求呢？”

“Well, no one seems to want a picture with me! People say I am vermin, when all I do is eat the overripe fruit no one bothers to harvest…”

“The thousands of tourists arriving on our tiny islands do worry me. What will they eat? Frozen chicken, flown in? And how to satisfy their need for fresh water, and for power, so they can have the comfort of cool air?”

……看起来没人想跟我合影！

... no one seems to want a picture with me!

这儿没有足够的空间放置太阳能电池板。

There isn't enough space here for solar panels.

“事实上，这些嗡嗡作响的发动机让我在白天完全睡不着，在夜晚也倍感压力和疲劳。嗡嗡作响的风车也是一样。但是还有其他选择吗？人类必须戒掉对化石燃料的依赖成瘾！”

“这儿没有足够的空间放置太阳能电池板。而且也受够这些风车桅杆了！它们把深浅不一的华丽蓝色调构成的美丽海景给破坏了。”

“好吧，如果你在寻找解决方案的话，这是个能融合某些自然界力量的理想地点。就像悠悠球和风筝所利用的力量。”狐蝠说道。

“As it is, I can’t sleep during the day with all the humming generators, that also stress me out at night. Humming windmills do the same. But what other options are there? People have to kick their addiction to fossil fuels!”

“There isn’t enough space here for solar panels. And enough with all these windmill masts! They destroy the wonderful views, of seascapes with all the gorgeous shades of blue.”

“Well, if you are looking for a solution, this is the ideal spot right here to blend some forces of Nature. Like those used by the yo-yo and the kite,” Flying Fox says.

“悠悠球？还有风筝？你肯定是在开玩笑吧！”

“有时候如果你希望提出能真正改变世界的解决方案的话，你必须得从一些非常疯狂的主意开始……”

“据我所知，悠悠球利用重力滚落，并通过及时拉动绳子从而再度向上滚起。这跟风筝有什么关系？”

“A yo-yo? And a kite? You’ve got to be kidding me!”

“Sometimes you have to start with some pretty wild ideas if you want to come up with solutions that will truly change the world...”

“As far as I know, a yo-yo uses gravity to roll down, and a timely tug on the string to roll up again. What does that have to do with a kite?”

悠悠球？

A yo-yo?

……风筝……

… a kite …

“这么说吧，一旦风筝升到空中……”

“……风筝会一直飘着。我知道的！但那跟悠悠球有什么关系？”

“一旦悠悠球开始运动，在重力作用下就会一直运动。”

“Well, once a kite is in the air …”

“… It keeps on flying. I know! But what does that have to do with a yo-yo?”

“Once the yo-yo gets going, with the force of gravity, it will keep on going.”

“所以当你把这些力结合起来，就产生了永恒的动力。我明白了。哇，你真是个天才！尽管我承认我一点都不明白这个机制是怎么运行的。”

“这么说吧，你已经找到了在海里用空气捕捉磷虾的方法，而我学会了从树上拿到芒果。让工程师们去琢磨把悠悠球和风筝所利用的力量结合起来的方法，把这个想法从设想变成现实吧。”

“也许他们也可以利用布谷鸟钟的原理？”

“So, when you combine these forces, there is perpetual power. I get it. Wow you are a genius! Although I admit I have no clue how the mechanics work.”

“Look, you’ve figured out how to catch krill in the sea with air, and I’ve learnt how to get to mangos from a tree. Let the engineers figure out how to combine the forces a yo-yo and a kite use, to take this idea from vision to reality.”

”“Maybe they could use a cuckoo clock too?”

……布谷鸟钟。

... a cuckoo clock too.

耐心听我说。

Just bear with me.

“这个想法也太古怪了吧？”狐蝠惊呼道。

“耐心听我说……当那些链条上沉重的仿真松果开始下落后，布谷鸟钟可以运作多长时间？”

“它的钟摆会连续不断摆动8天。所以它在每次摆动过程中可以传递能量，在小布谷鸟冒出头来的时候还可以播放一小段旋律……这主意真太有用啦！”

“What for an outlandish idea is that?” Flying Fox exclaims.

“Just bear with me… For how long will a cuckoo clock keep ticking once those weights, those fake pine cones on chains, start moving downward?”

“Its pendulum will swing for eight days – non-stop. So it can deliver power, for a tick and a tock, and a little melody when the birdie pops out… A very useful idea, indeed!”

“谢谢。现在到了我们在马尔代夫开始创新，提供我们自己的解决方案的时候了。毕竟我们可是与众不同的环礁国家。”

“很快我们就会以‘风筝之国’而著称——由悠悠球和布谷鸟钟技术供能。这会吸引全世界的注意的！”

……这仅仅是开始！……

“Thanks. It’s about time we in the Maldives started innovating, to provide our own solutions. We are, after all, an atoll nation like no other.”

“And soon we will be known as ‘the kite nation’ – powered by yo-yo and cuckoo clock technology. That will get the world’s attention!”

… AND IT HAS ONLY JUST BEGUN!…

……这仅仅是开始！……

… AND IT HAS ONLY JUST BEGUN! …

Did You Know?

你知道吗？

An atoll is a ring-shaped coral reef with a lagoon. The word 'atoll' is from the Dhiveni language, only spoken in the Maldives, a nation made up of 1,200 islands and atolls.

环礁指的是围绕潟湖的环状珊瑚礁。“环礁（atoll）”一词来自迪维希语，这种语言只有在马尔代夫有人使用，马尔代夫是一个由1 200座岛屿和环礁组成的国家。

The yo-yo was first used 2,500 years ago, by Greek youths whomastered the timely pull of the string, to make images of the gods painted on the yo-yo rotate as the yo-yo goes up and down. Playing with a yo-yo helps one to focus and control temper.

早在2 500年前，希腊的青年开始使用悠悠球，他们掌握了及时拉绳的技巧，使溜溜球上所绘的神像在悠悠球上下运动时旋转起来。玩悠悠球有助于集中注意力和控制脾气。

“悠悠球”一词来自菲律宾吕宋岛北部所使用的伊洛卡诺语。一个来自菲律宾的男人在美国加州的圣巴巴拉开了第一家悠悠球店。在1929年，600名员工每天可以生产30万个悠悠球。

The word ‘yo-yo’ originates from the Ilocano language spoken in Northern Luzon, Philippines. A man from the Philippines opened the first yo-yo shop in Santa Barbara, California (USA). In 1929, a staff of 600 produced 300,000 units per day.

风筝起源于亚太地区，最早可追溯到9 000多年前的苏拉威西（印度尼西亚）。大约2 500年前，新西兰和中国发明了自己的风筝。

The kite originated in the Asian Pacific Rim, with first traces of it found in Sulawesi (Indonesia) more than 9,000 years ago. New Zealand and China invented their own versions around 2,500 years ago.

The Wright brothers developed man-lifting kites, as precursors to the first airplane. The advent of mechanically powered aircraft diminished the interest in kites, which eventually re-emerged as hang gliders.

莱特兄弟发明了载人风筝，作为第一架飞机的前身。机械动力飞机的出现降低了人们对风筝的兴趣，风筝最终以悬挂式滑翔机的形式再现。

The cuckoo clock was first manufactured in the 1600s, in the Black Forest of Germany. It is a German company, SkySails, that has now integrated kite, yo-yo and clock technologies into the most efficient kite-power technology.

布谷鸟钟最早于17世纪在德国黑森林被制造出来。一家名为天帆（SkySails）的德国公司目前已将风筝、悠悠球和布谷鸟钟技术整合为最高效的风筝发电技术。

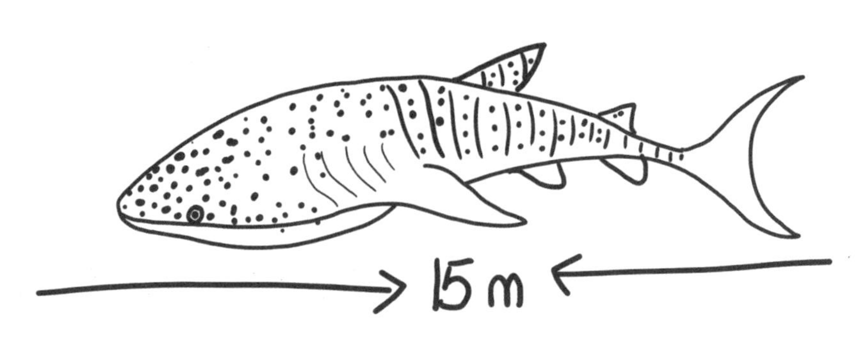

The whale shark is not a whale but a fish, measuring up to 15 metres in length, and living for up to 100 years. It filter-feeds on plankton and tiny fish. Female whale sharks (like some bats) store the male's sperm and control when fertilisation will take place.

鲸鲨不是鲸，而是一种鱼类，长达15米，寿命可长达100年。它以浮游生物和小鱼为食。雌性鲸鲨（像一些蝙蝠一样）储存雄性鲸鲨的精子并控制受精时间。

The flying fox is not a fox, but is a highly social bat, with up to 500 of them living in the same tree, feeding at night and sleeping in an upside-down position during the day. The highest ranked bats live at the top of the tree.

狐蝠不是狐狸，而是一种高度群居的蝙蝠，在同一棵树上生活的狐蝠可以高达500只，它们夜晚觅食，白天以倒吊的姿势睡眠。地位最高的蝙蝠生活在树的最顶端。

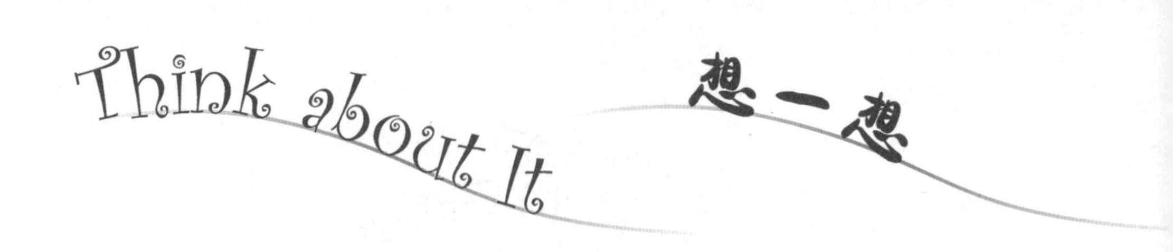

Could you ever have imagined that the forces that make a kite and a yo-yo work would be integrated?

你能想象使风筝和悠悠球运动的力量能被整合到一起吗?

Is the cuckoo clock a mechanical system of the past?

布谷鸟钟是过时的机械系统吗?

Would you prefer using windmills in a lagoon to floating solar panels in the sea, or use neither?

比起把太阳能电池板放到海水里，你会更情愿在潟湖里使用风车吗？还是两者都不用?

Should solutions be the same everywhere in the world?

世界各地的解决方案应该是一样的吗?

Do It Yourself!

If there is little land space and there is no electricity, how would you solve the problem of generating power? Let's make an inventory of all the sources of energy available that do not need a lot of space. Add to this, we do not want the view blocked by windmill masts, and we want a source of energy that is local. List all the energy sources you can imagine that fit these criteria. What possible power sources come to mind, whether commercially available yet, or not? Discuss your ideas with your friends and family members and get their input and feedback. Now, use the three most inspiring ideas to envision your dream solution.

如果可利用的土地面积很少且没有电，你会如何解决发电难题？让我们把所有对空间要求不高的可用能源列一个清单。再加上这些条件——不让风车桅杆挡住风景，使用本地能源。列出所有你能想到的符合标准的能源。如果不考虑经济上是否可行，你能想到哪些可能的能源？与你的朋友和家人讨论，并获得他们的意见反馈。现在，用三个最令人振奋的想法来设想你的解决方案。

学科知识

Academic Knowledge

生物学	印度狐蝠是一种果蝠，也是世界上最大的蝙蝠之一；狐蝠对人类来说是疾病传播媒介；榕树70%的种子是由狐蝠传播的；鲸鲨是滤食性动物和世界上最大的脊椎动物；鲸鲨生活在热带海区的中上层；鲸鲨是卵胎生，在体内进行产卵和孵化；环礁的形成是由于海洋有机物生长。
化　学	氮磷循环是环礁形成有机过程的一部分。
物　理	悠悠球通过摆动和旋转将势能转换为两种动能，通过拉动绳子产生了能量；质量和速度的结合产生了动量；风筝结合重力、阻力、推力和升力；重量通过引力驱动钟摆来回摆动，每次摆动需要一秒钟；随着钟摆的来回摆动，势能和动能不断转换。
工程学	陀螺仪沿着三个坐标轴的角动量分量都守恒；擒纵器是一种利用重力实现计时目的的齿轮装置。
经济学	巴西在养鸡业的竞争地位来自动物养殖和饲料生产的纵向一体化；当使用本地能源并且不进口石油的时候，巴西的经济和社会增长潜力巨大，因此原本用来购买石油的资金现在得以留在巴西国内以刺激增长。
伦理学	与当地负担不起水电的人相比，旅游者更容易用上水电，然而旅游业一般无法靠提供低收入工作使当地人受益；自由贸易以高昂的能源成本为偏远地区带来低质量的食品，阻止当地社区增强其复原力以及确保粮食安全，使其依赖进口；人类把只吃过熟的水果和作为授粉者的蝙蝠归类为害兽。
历　史	伽利略研究过钟摆的性质，由于钟摆来回摆动的时间是相同的，因此这些钟摆取代了以前所有的时钟机制；查尔斯·达尔文经过五年的观察，在1842年解释了环礁的成因；第一个布谷鸟钟制造于17世纪。
地　理	热带水域的环礁或环状珊瑚礁；大多数环礁位于太平洋和印度洋；马尔代夫是最大的环礁国家。
数　学	多能源集成设计需要用到人工智能；计算位置、速度和加速度。
生活方式	度假，体验新奇和不同的事物；依赖他人寻找解决方案并复制模仿，或者寻找那些考虑独特条件的解决方案。
社会学	为自己拍照（自拍）的习惯；布谷鸟钟作为天真、童年、老年、过去、惊讶和快乐时光的隐喻。
心理学	永久性背景噪音引起压力水平提高；笑话和笑声释放压力；利用头脑风暴想出点子。
系统论	狐蝠种群的衰退是由于人们相信它们的脂肪是治疗风湿病和哮喘的良药；大众旅游的影响和游客数量过多造成的损害；风筝发电很快就可以试行，而且可以在没有任何排放的情况下产生电力。

情感智慧

Emotional Intelligence

蝙　蝠

蝙蝠担心游客在鲸鲨栖息地附近乘船疾驶。人们认为他是害兽使他很伤心。他无力改变因噪音污染而无法入睡的困境。他提出了一个惊人的想法，并坚称疯狂的构想将解决问题。想法的灵感来自对自然界力量的观察，他也知道自己的局限性。他认为专家们可以解决工程方面的问题。鲸鲨提出的使用布谷鸟钟机械原理的建议出乎蝙蝠的意料。他很快恢复镇静并开始分析，展示了他的创造性思维。用对副作用的反思来调和自己的想法，表明他确实具有系统性思维。

鲸　鲨

鲸鲨对找到理想的生活场所感到满足。她抱怨那些只关注自拍的游客，以及大众旅游对环境造成的伤害。起初她不相信蝙蝠的建议，直言看不到想法间的联系，要求对方做出解释。她在头脑风暴过程中提到的一个古怪想法给蝙蝠带来了惊喜。随着自信心增强，她给出了更多的细节。她积极渴望为共同利益做贡献。她为自己生活的环礁国家感到自豪，并鼓励马尔代夫人民找到自己的解决方案，而不是效仿在其他地方已实施的办法。

艺术

The Arts

让我们在天空中创造风筝“跳舞”的艺术。你可以拉动风筝线让风筝在室外移动，还可以让风筝在室内飞动——控制两根线就能让风筝在没有风的时候也能停在空中。这是一门艺术：用自己扇动的风来放风筝。用不同的手臂做动作，以不同的方式走动，创造各种有趣的风筝轨迹。无论是在室内还是室外，用风筝在空中“作画”都是在用升力克服重力，用推力克服阻力的情况下进行的。这些力量会让你的风筝继续飞翔和移动，创造移动的艺术！

思维拓展

Systems: Making the Connections

过度依赖化石燃料是气候变化的元凶之一，现有可再生能源也存在弊端，如风车的视觉和噪音污染，或者太阳能电池板所需的空间。岛国马尔代夫需找到因地制宜的解决方案。通过人工智能对气候数据的智能处理来运行风筝的转型受到欢迎。这些风筝每天24小时连续工作，还可以自动启动。由于磨损影响，风筝的纺织材料需不时更换。由人工智能管理的风筝可在空中停留数月，不断寻找利用风力的最佳高度，并在飞机靠近时调整风筝的高度。这种极具独创性的想法把驱动悠悠球和风筝的力量结合起来，为偏远地区的人们提供动力解决方案。只要安装了本地配电网，风筝系统就可以在几分钟内与配电网连通。使用风筝作为主要的发电系统，不仅价格低廉，还可改变一个岛屿的形象。据计算，马尔代夫只需要250只这样的风筝就可以实现电力的自给自足。随着智能电网的出现，这样的风筝将成为一道独特的风景。孩子们可以在度假的同时了解利用风力发电的未来。

动手能力

Capacity to Implement

让我们来谈谈制作风筝的核心元素，主要目的是学习如何设计和制造不同类型、具有不同性能的风筝。首先，找到方向，然后制作一个只有一根绳子的风筝。你没办法操控这样的风筝做很多不同的移动。所以，一旦你有了单绳风筝的经验，你可以开始做第二个风筝——更复杂的双绳风筝。即使没有风的时候，仅仅通过操纵风筝上的两根绳就可以让它飞起来。制造风筝是一种已经存在了几千年的传统。学习如何制作你自己的风筝可能需要时间，因为你首先必须了解基本知识。这么多年来很多人已经掌握了制作方法。你们会吗?

故事灵感来自

This Fable Is Inspired by

斯蒂芬·雷格

Stephan Wrage

斯蒂芬·雷格出生于德国汉堡，爱好航行和风筝冲浪。他毕业于德累斯顿理工大学，获得了商业工程学位，主修机械制造、物流、创新管理和控制。作为工程训练的一部分，他在一个团队里工作并建立了一个新的电机厂。斯蒂芬一直想追求以风筝为中心的愿景，于是在2001年创办了天帆公司。它为航海船只配备帆。2009年，天帆公司为首批6艘货船装备了风筝。把静态风筝变成动态风筝的想法很快产生了。这一想法把悠悠球和人工智能的逻辑与布谷鸟钟的机制结合了起来。该系统已经被证明是可行的，为能源供应和向内陆地区输送电力提供了一个实用的选择。印度洋地区宣称在这一领域处于领先地位。

图书在版编目(CIP)数据

冈特生态童书.第七辑：全36册：汉英对照 /
(比)冈特·鲍利著；(哥伦)凯瑟琳娜·巴赫绘；
何家振等译.—上海：上海远东出版社，2020
ISBN 978-7-5476-1671-0

Ⅰ.①冈… Ⅱ.①冈… ②凯… ③何… Ⅲ.①生态
环境-环境保护-儿童读物—汉英 Ⅳ.①X171.1-49

中国版本图书馆CIP数据核字(2020)第236911号

策　　划 张　蓉
责任编辑 程云琦
助理编辑 刘思敏
封面设计 魏　来 李　廉

冈特生态童书

悠悠球和风筝

[比]冈特·鲍利　著
[哥伦]凯瑟琳娜·巴赫　绘
贾龙智子　译